First Library of Knowledge

How Things Work

BLACKBIRCH PRESS

An imprint of Thomson Gale, a part of The Thomson Corporation

THOMSON

GALE

Detroit • New York • San Francisco • San Diego • New Haven, Conn. • Waterville, Maine • London • Munich

First published in 2005 by Orpheus Books Ltd., 2 Church Green, Witney, Oxfordshire, OX28 4AW

First published in North America in 2006 by Thomson Gale

Copyright © 2005 Orpheus Books Ltd.

Created and produced: Rachel Coombs, Nicholas Harris, Sarah Harrison, Sarah Hartley, Emma Helbrough, Orpheus Books Ltd.

Text: Nicholas Harris

Consultant: Chris Oxlade

Illustrator: Mike Fuller

For more information, contact
Blackbirch Press
27500 Drake Rd.
Farmington Hills, MI 48331-3535
Or you can visit our Internet site at http://www.gale.com

LIBRARY OF CONGRESS CATALOGING-IN-PUBLICATIONS

Harris, Nicholas, 1956-
   How things work / By Nicholas Harris.
      p. cm. -- (First library of knowledge)
   Includes bibliographical references and index.
   ISBN 1-4103-0346-2 (hardcover : alk. paper) 1. Machinery--Juvenile literature. I. Title. II. Series.

TJ147.H245 2006
600--dc22   2005029619

Printed in Malaysia
10 9 8 7 6 5 4 3 2 1

CONTENTS

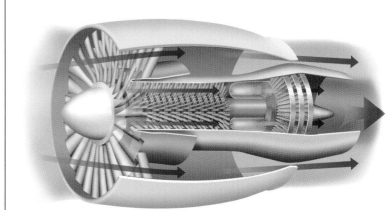

INTRODUCTION

WHAT would life be like without machines? It would be difficult to travel very far except on horseback. It would take much longer to make things. And the world would be very different without telephones, televisions, or computers. Modern machines make our lives easier and more enjoyable.

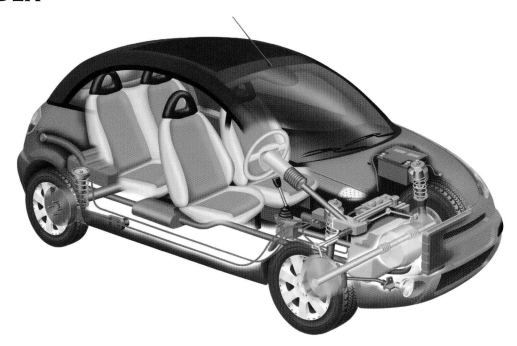

HEAVY VEHICLES

A BULLDOZER uses its blade to push earth and rubble. A digger can tear into the ground using the teeth on its bucket. It scoops up the rubble and pours it into the back of a dump truck. The truck has large wheels and a powerful engine. This enables it to carry very heavy loads. To dump a load, the truck body tips up, the tailgate at the back swings open, and the rubble slides out.

Toothed bucket

Bulldozer

Tailgate

Truck body

Blade

CATERPILLAR TRACKS

On this demolition site, many heavy vehicles have caterpillar tracks instead of wheels. They allow the machines to move over soft or uneven ground. A crane swings a heavy wrecking ball into the walls of the old buildings. This sends the bricks crashing to the ground.

Jib

Wrecking ball

Digger

Crane

Dump truck

CONSTRUCTION MACHINES

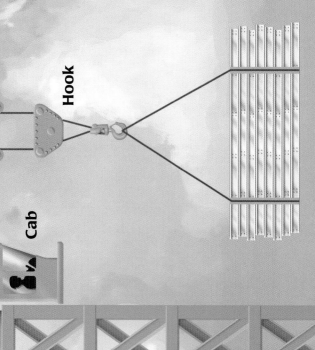

Jib

Hook

Cab

Counter-weight

Tower

MANY machines are used in the construction of a tall building. A tower crane is a vital machine. It lifts heavy building materials into place. Here, it lifts steel girders to the upper floors. These will make up the building's frame. Concrete slabs, which form the floors and walls, will then be lifted by the crane.

The hook, which carries the load, moves along the crane arm, or jib.

The driver operates the crane from the cab. A heavy concrete counter-weight at the back of the crane balances the weight of the load being lifted by the crane.

CONCRETE

Concrete is made from cement, sand, crushed rock, and water. The drum of the concrete mixer spins, mixing the ingredients together. It forces the liquid concrete out of the back of the truck.

Pumping concrete into the foundations

Building under construction

Crane base

Concrete mixer

CARS

CARS are powered by engines fueled by gasoline or **diesel**. The engine supplies energy to turn the wheels. The driver can go faster by pressing the **accelerator** pedal. This increases the flow of fuel to the engine. Gears allow the car to be driven at different speeds without the engine running too slowly or too quickly. The suspension system uses springs to give a comfortable ride.

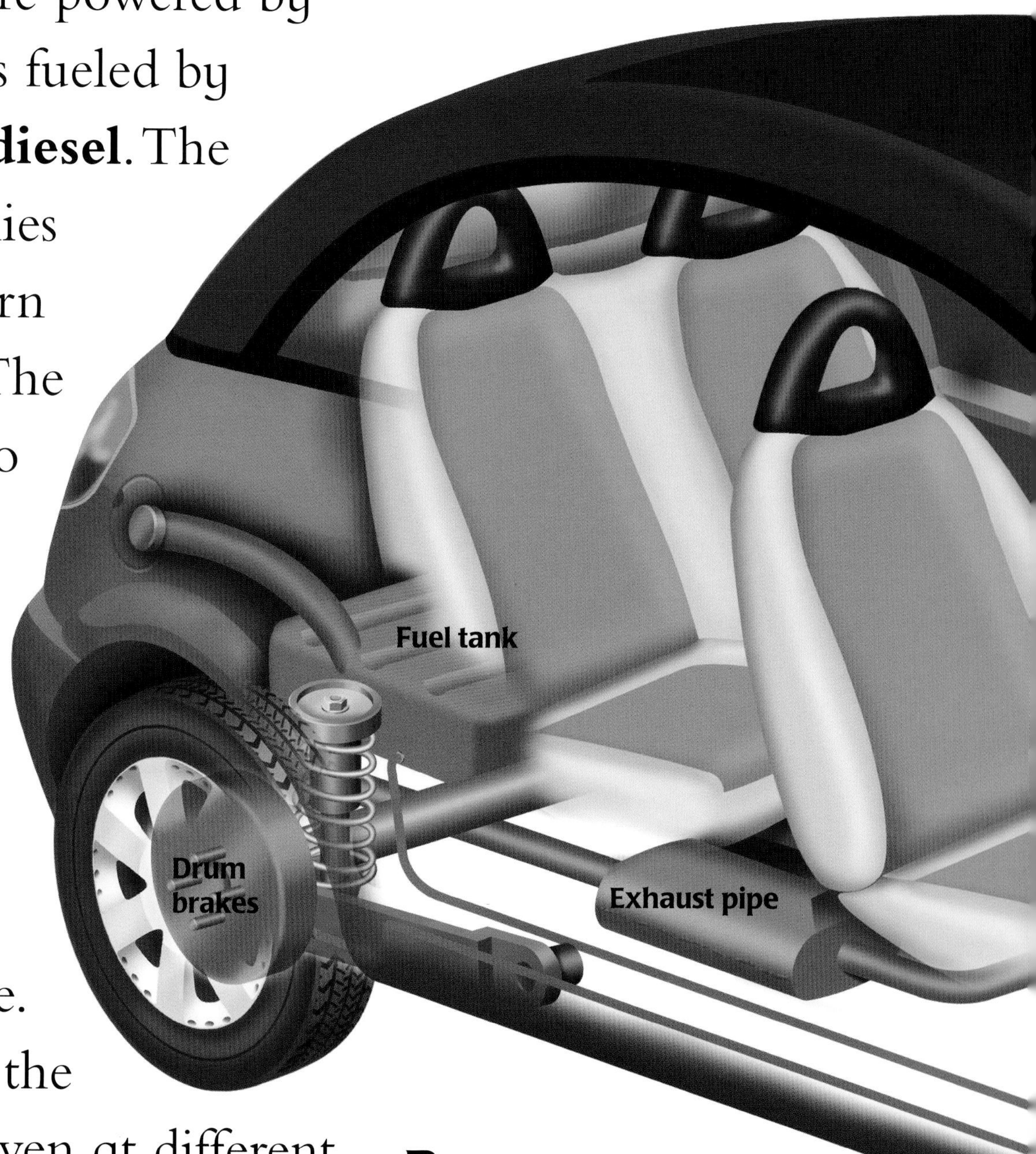

Fuel tank

Drum brakes

Exhaust pipe

BRAKES

To slow down, the driver presses the brake pedal. This causes brake pads to squeeze a disc inside the wheels. A hand brake operates brake pads inside drums attached to the rear wheels.

A battery supplies electricity to a starter motor. This starts the engine.

GASOLINE ENGINE

A gasoline engine works by burning a mixture of gasoline and air. Gasoline is drawn from the fuel tank and sprayed into the engine's cylinders. There it is ignited (set on fire) and turned into hot gases. The hot gases force the pistons down. This action turns a crankshaft that is connected to the wheels.

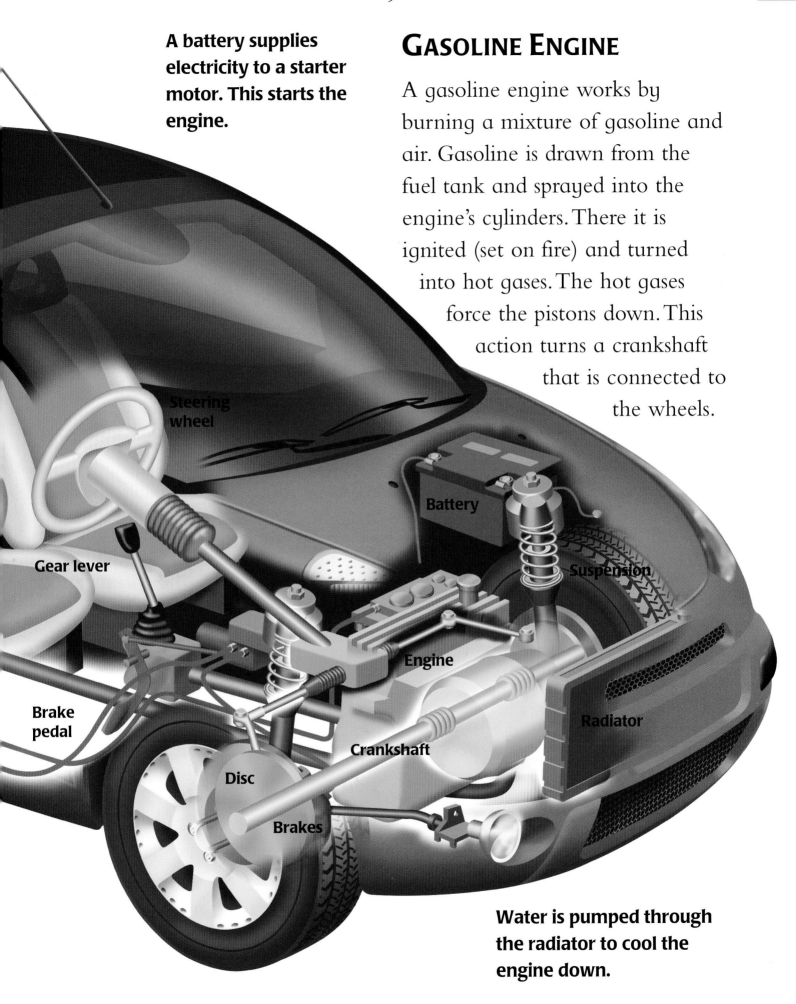

Steering wheel

Battery

Gear lever

Suspension

Engine

Brake pedal

Radiator

Crankshaft

Disc

Brakes

Water is pumped through the radiator to cool the engine down.

TRAINS

HIGH-SPEED trains are powered by electric motors. In diesel- electric trains, a diesel engine drives a generator, producing the electricity to power the motors.

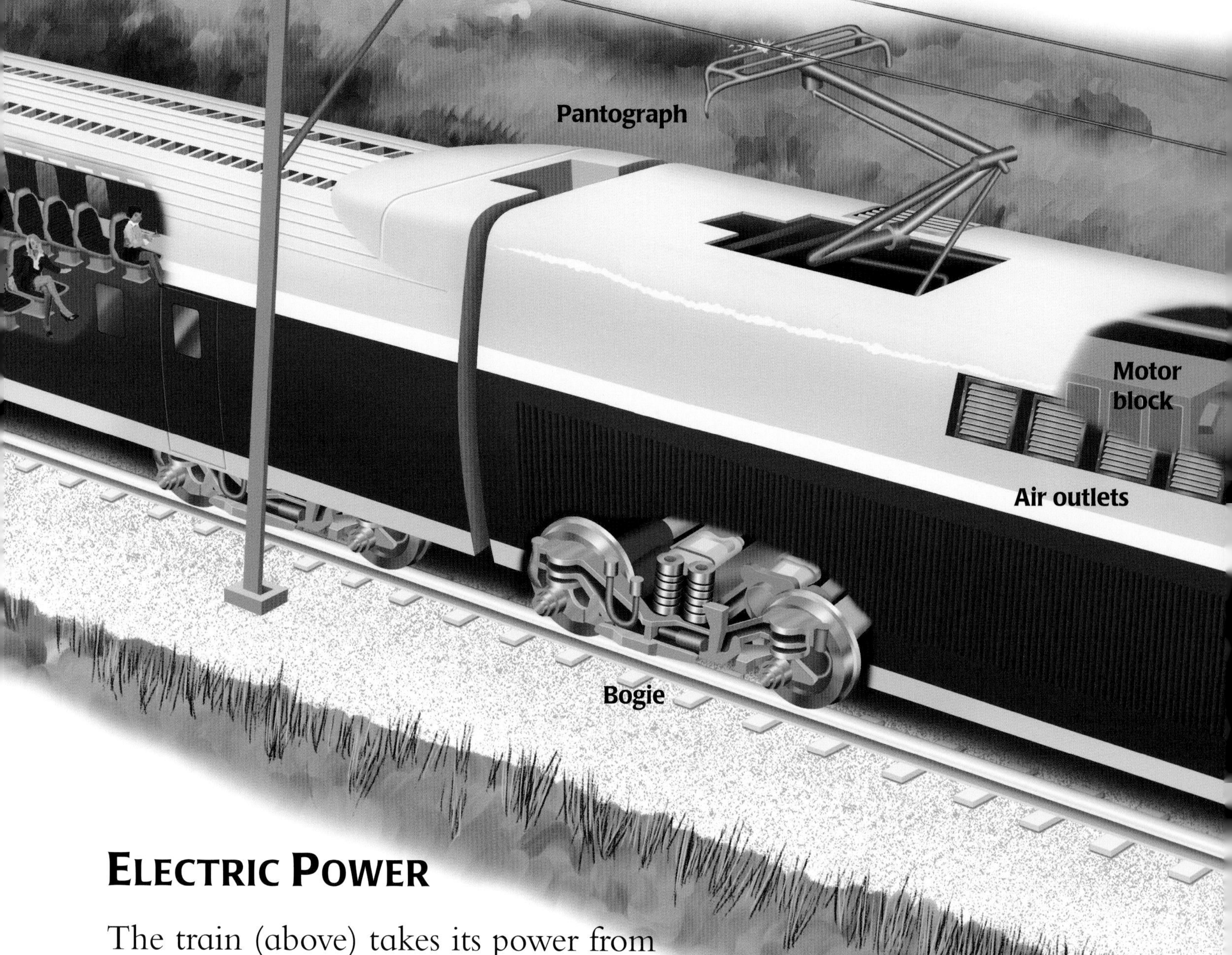

Pantograph

Motor block

Air outlets

Bogie

ELECTRIC POWER

The train (above) takes its power from electricity cables running above the tracks. A pantograph collects the electricity. Then a motor block and a **transformer** control the flow of electricity to the motors.

MAGNETIC TRAINS

Maglevs, short for magnetic levitation, are trains that use magnets to hover above the rails. This removes the friction from wheels on rails that limits the speed of ordinary trains. Maglev trains can therefore travel very fast. They have been tested at speeds of up to 340 miles (550 kilometers) per hour. Maglevs also use much less fuel.

Electricity cables

Cooling fans

Transformer

Air conditioning unit

Driver's cab

Bogies are made up of wheels, brakes, gears, motors, and suspension.

SHIPS AND BOATS

SHIPS are large vessels that can travel across the sea. Boats are smaller craft. Most ships are powered by engines that turn a propeller.

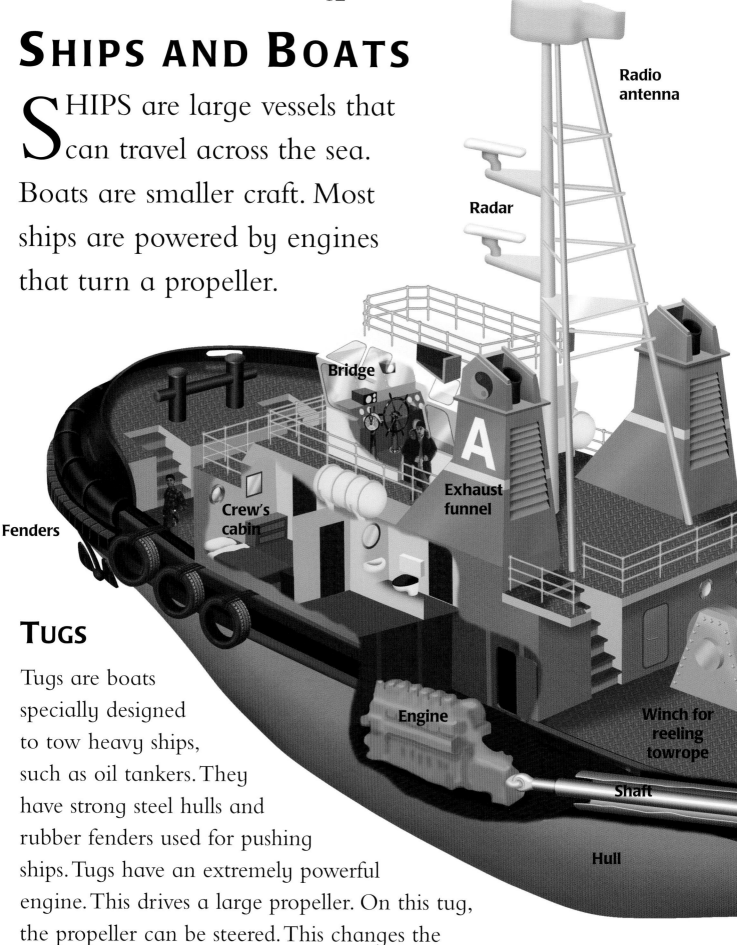

Radio antenna

Radar

Bridge

Exhaust funnel

Fenders

Crew's cabin

Winch for reeling towrope

Engine

Shaft

Hull

TUGS

Tugs are boats specially designed to tow heavy ships, such as oil tankers. They have strong steel hulls and rubber fenders used for pushing ships. Tugs have an extremely powerful engine. This drives a large propeller. On this tug, the propeller can be steered. This changes the direction that the tug is moving in.

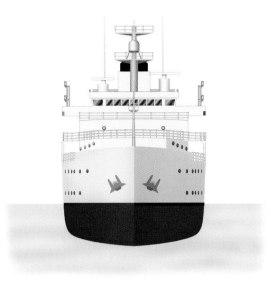

WHY A SHIP FLOATS

Water pushes upward on anything that is immersed in it. Anything that is denser than water will sink. Objects less dense than water will float. A heavy metal ship floats because it contains lots of air. This makes it less dense than water.

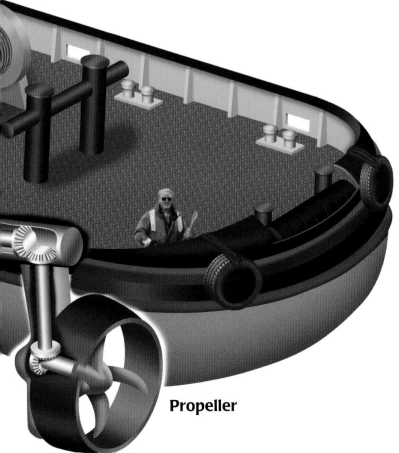

Propeller

SUBMARINES

Submarines are vessels that can travel underwater. To dive, water is let into special hollow spaces called ballast tanks, making the submarine heavier. To surface, the water is pumped out of the tanks.

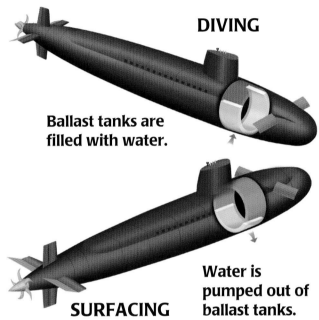

DIVING

Ballast tanks are filled with water.

Water is pumped out of ballast tanks.

SURFACING

PROPELLER

Like some airplanes, many ships and boats are driven by a propeller. It is connected to the engine by a shaft. The propeller's blades are large and curved. As they turn, the water around them is sucked in and pushed backward. This drives the boat forward.

AIRCRAFT

ALL flying machines are types of aircraft. Airplanes and helicopters are heavier than air. They need wings or spinning blades and engines to keep them in the air. Balloons and airships stay airborne because they are filled with lighter-than-air gas.

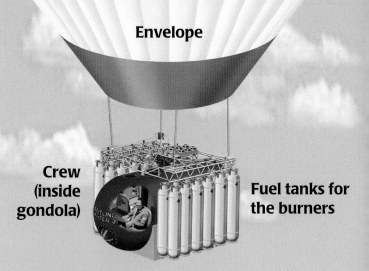

Envelope

Crew (inside gondola)

Fuel tanks for the burners

BALLOON

Hot-air balloons have a large bag called an envelope that contains air. Burners heat the air in the envelope from below, creating lift. The crew travels in a gondola attached beneath the envelope.

Wing

Passenger cabin

Baggage hold

Flight deck

Engine

In this drawing, parts of an airliner have been cut away to show the inside.

HELICOPTERS

Most helicopters are powered by a turboshaft engine—a jet engine that turns a shaft, causing the rotor blades to spin very quickly. They push the air downward, which lifts the aircraft upward.

Helicopters can take off and land vertically. They can hover and fly in any direction.

Tail fin

Tailplane

Most airplanes have a central tube called a fuselage. The tailplane and fin help keep the plane flying straight and level. Some modern airliners can carry more than 500 passengers.

JET ENGINE

In a jet engine, a giant fan sucks in air. Some of this air passes through compressors that raise the air pressure. Burning fuel produces hot gases that mix with the rest of the air as it blasts out the back of the engine. The force drives the engine (and the airplane, that is attached to it) forward.

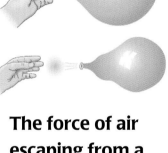

The force of air escaping from a balloon sends it in the opposite direction. This is how a jet engine works.

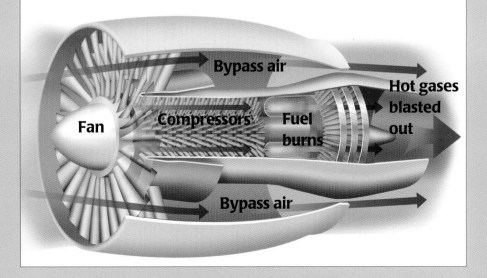

Bypass air

Hot gases blasted out

Fan

Compressors

Fuel burns

Bypass air

HOW A PLANE FLIES

A HEAVY airplane is able to fly because of the shape of its wings. When the plane is driven forward at high speed, a force called lift causes it to take off.

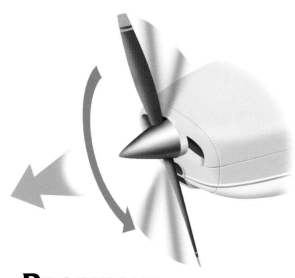

PROPELLER

Some planes are powered by propellers. Each blade has a curved shape. As the propeller turns, the blades suck in air and sends it backward. This pulls the plane forward.

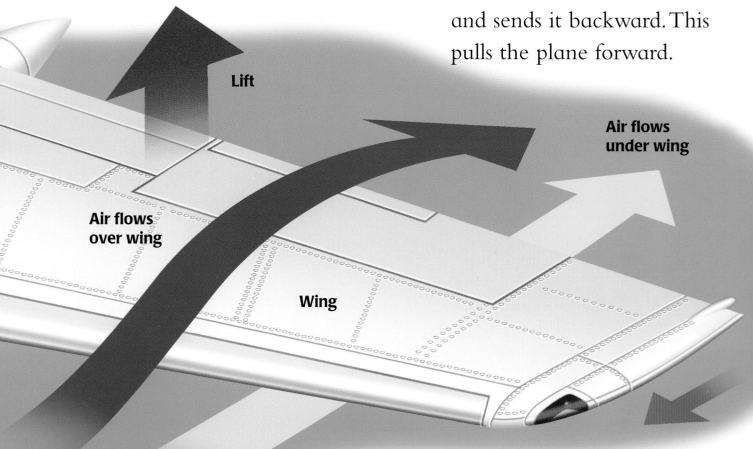

Lift

Air flows under wing

Air flows over wing

Wing

Because of a wing's curved shape, the air flowing over it moves faster than that flowing under it as the plane moves forward. The air flowing over the wing has lower pressure. An upward force, called lift, is the result.

STEERING

An airplane moves in three different ways. It can climb or dive (pitching). It can turn to the left or right (yawing). And it can bank to one side or the other (rolling).

Nose up   **PITCHING**

Elevator

Tail down

To cause an airplane to climb, the pilot pulls the control column back. This raises flaps called elevators on the tail. The airflow is turned upward, causing the airplane's tail to drop and its nose to rise. To make the aircraft dive, the pilot pushes the control column forward. This has the opposite effect on the airflow. The tail rises and the nose drops.

Nose right

Rudder

YAWING

Tail left

To turn the airplane to the right or left, the pilot uses pedals to swivel the rudder on the tailfin. This changes the airflow. To turn the aircraft smoothly, the pilot must also roll the aircraft. Raising a flap (an aileron) on one of the main wings causes the plane to roll to one side.

Wing up

Aileron

Aileron

ROLLING

Wing down

SPACE ROCKETS

TO TRAVEL into space, a space rocket must reach a speed of 25,854 miles (40,000 kilometers) per hour. This is the minimum speed needed to escape the Earth's **gravity**. Rocket engines must therefore be both very powerful and able to work without air (there is none in space).

In a rocket engine, two different fuels are mixed together. They make hot gases that rush out through a nozzle at great speed. This propels the rocket upward.

A space rocket is made up of separate stages. When the fuel in one stage is burned up, that stage is jettisoned (cast off into space).

A satellite to be launched into space (the "payload")

Liquid hydrogen fuel tank

Liquid oxygen fuel tank

Rocket engine (second stage)

Liquid oxygen fuel tank

Kerosene fuel tank

Solid-fuel booster engine

Solid-fuel booster engine

Rocket engine (first stage)

Space Probes

It would take a manned spacecraft too many years to reach distant planets. To explore worlds such as Jupiter and Saturn, remote-controlled space probes have been launched instead. Fired into space by rockets, they need no engines since there is no air to slow them down. They send back photos of the planets as they fly near them.

The space probe *Cassini* went into orbit around Saturn in 2004. Then a lander was parachuted down to Saturn's moon Titan.

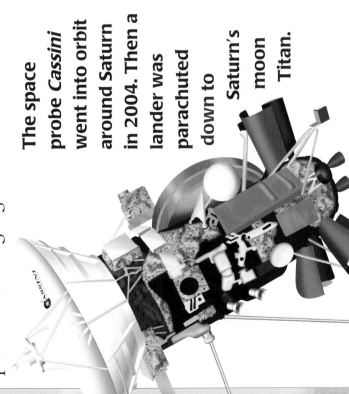

TELESCOPES

A TELESCOPE makes things that are far away appear much larger. It can show details that are not visible to the naked eye. In a reflecting telescope, light is reflected by a dish-shaped primary mirror onto a smaller secondary mirror, and from there to the viewer or light sensor.

Light from object

Secondary mirror

Image reflected onto light sensor

Primary mirror

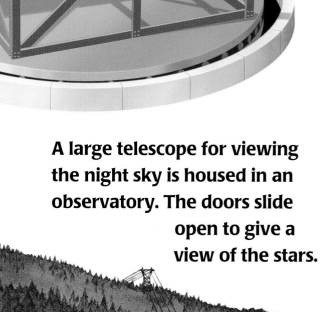

A large telescope for viewing the night sky is housed in an observatory. The doors slide open to give a view of the stars.

OBSERVING SPACE

For astronomers, scientists who study space, telescopes are essential. A powerful reflector reveals details of planets, nebulae (clouds of dust or gas), and galaxies that are invisible to the naked eye. This view *(right)* of a galaxy billions of miles away was taken by the Hubble Space Telescope.

SPACE TELESCOPE

Orbiting 385 miles (620 kilometers) above Earth is the Hubble Space Telescope. Since there is no air in space, it can see objects much more clearly than telescopes on Earth are able to do. It is so sensitive, it can detect light from a flashlight 248,548 miles (400,000 kilometers) away.

Cover

Images are sent to Earth through the space telescope's antennae.

Antenna

Primary mirror

Secondary mirror

Solar panel

Sensors

Antenna

This illustration shows the mirrors inside the Hubble Space Telescope.

CAMERAS

CAMERAS are used for taking photographs. Ordinary cameras record images on photographic film. Digital cameras record images electronically.

DIGITAL CAMERA

Light from a scene is let into the camera when the shutter opens. The lens focuses light to form an image on a **microchip**. This divides the image into thousands of tiny **pixels**. The chip records the color of each pixel in its memory. The image can then be transferred to a computer.

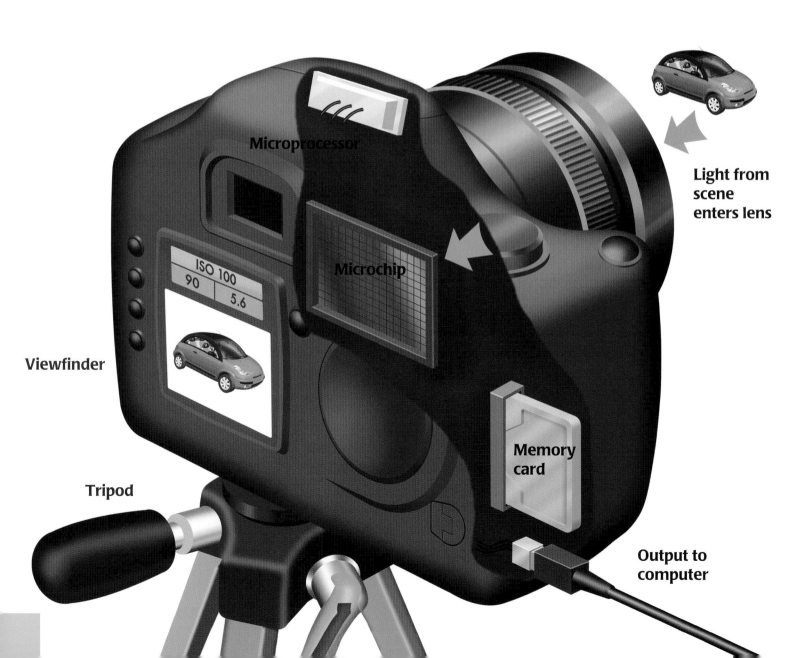

Microprocessor

Microchip

ISO 100
90 | 5.6

Light from scene enters lens

Viewfinder

Memory card

Tripod

Output to computer

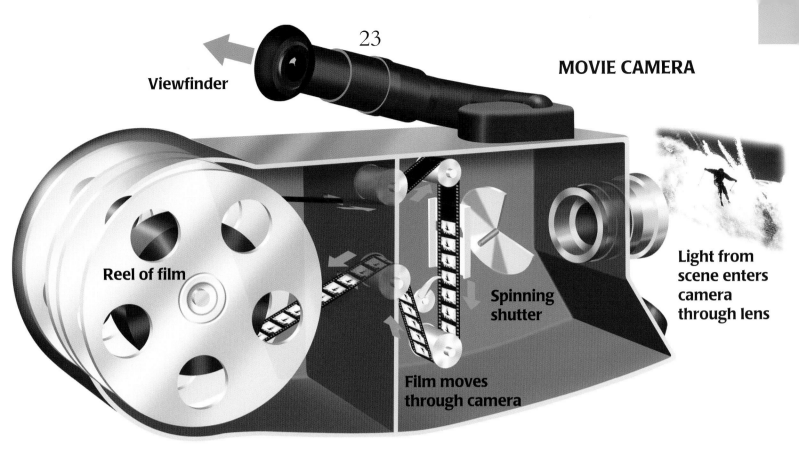

MOVIE CAMERA

Viewfinder

Reel of film

Spinning shutter

Film moves through camera

Light from scene enters camera through lens

MOVIE CAMERA

A movie camera takes thousands of photographs, called frames, of a moving scene. For each second of action, 24 frames are stored on film. Inside the camera is a spinning shutter. It opens to let light onto the film. This creates a frame. Then it closes and the film is moved into a new position for the next frame. This happens again and again until a whole reel of film is made.

A movie projector shows the film. A bright light shines through each frame and projects the image onto a screen. While the shutter is closed, the film moves on a frame. The frames pass through so quickly that the image appears to move.

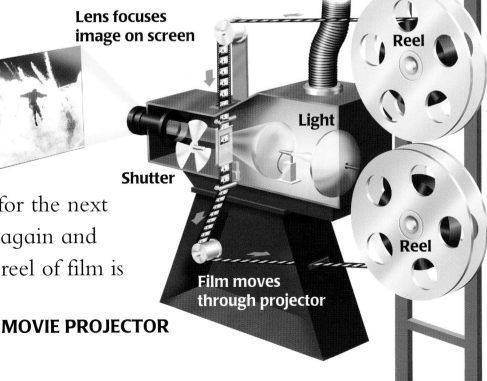

Lens focuses image on screen

Reel

Light

Shutter

Reel

Film moves through projector

MOVIE PROJECTOR

DIGITAL RECORDING

INFORMATION can be recorded in digital form on a compact disc (CD). Sounds, images, video, or text are turned into a series of numbers represented by the digits 0 and 1. These are stored on the disc as a series of pits and flats. The disc is "read" by a laser beam.

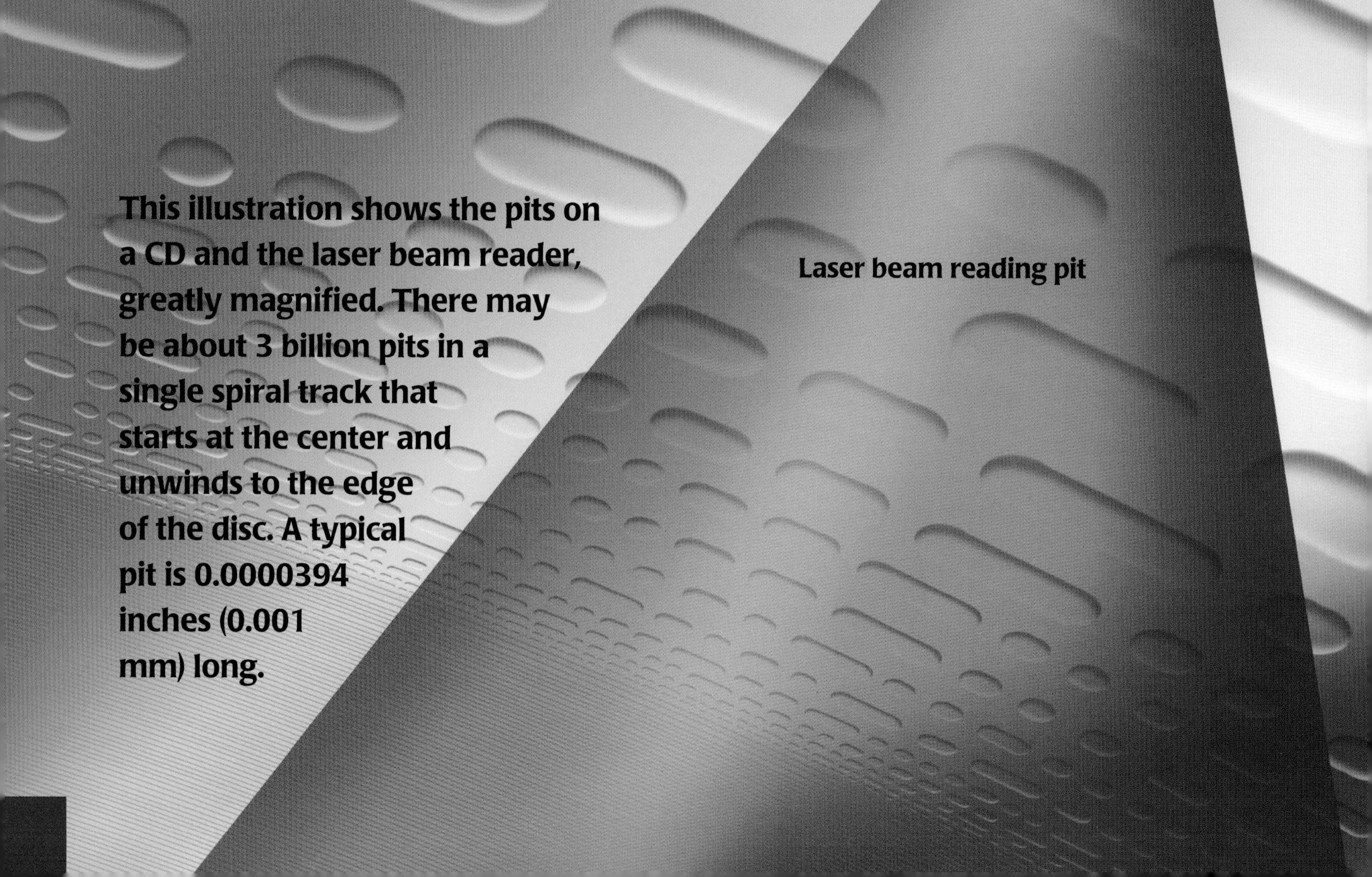

This illustration shows the pits on a CD and the laser beam reader, greatly magnified. There may be about 3 billion pits in a single spiral track that starts at the center and unwinds to the edge of the disc. A typical pit is 0.0000394 inches (0.001 mm) long.

Laser beam reading pit

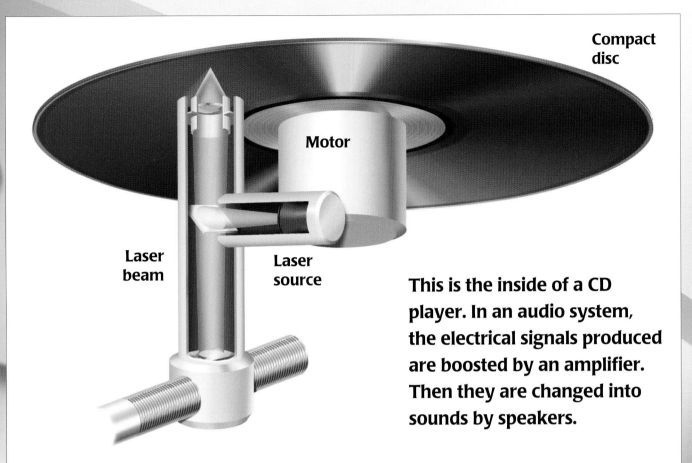

Compact disc

Motor

Laser beam

Laser source

This is the inside of a CD player. In an audio system, the electrical signals produced are boosted by an amplifier. Then they are changed into sounds by speakers.

COMPACT DISC PLAYER

In a CD audio player, the disc is read by a beam of laser light. A series of mirrors and lenses focuses the beam onto the underside of the disc. The reflected beam shows up the various pits and flats. The machine turns these into electrical signals. Many millions of electrical signals together make up a piece of music. Computer CD-ROMS holding data, text, or graphics and DVDs containing a movie or TV program, work in the same way.

Pit

Flat

A compact disc (CD) and digital video disc (DVD) are both made of plastic and aluminium and measure 4.7 inches (12 centimeters) in diameter.

TELEVISION

Ground station

TELEVISION is a way of sending moving pictures from one place to another. There are thousands of TV channels around the world. Most programs are recorded, but some, especially news and sports, are broadcast live. Viewers see the action as it happens.

To make a TV program, a camera shoots the pictures while a microphone picks up sounds *(bottom left)*. Pictures and sound are combined in the control room *(below)*.

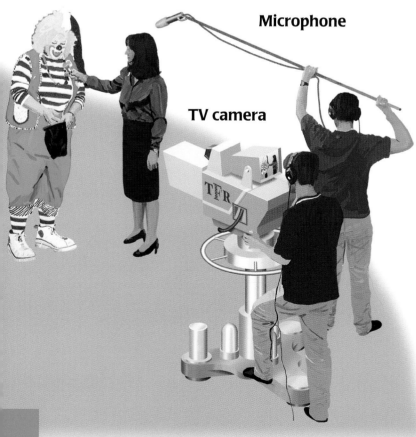

Microphone

TV camera

TELEVISION CAMERA

A TV camera contains **charge-coupled devices** (CCDs). As the camera shoots a scene, the CCDs collect light on millions of tiny squares called pixels. Each pixel measures the brightness and color of that light (made up of red, blue, or green) in digital form. The camera shoots the changing scene 25 times a second.

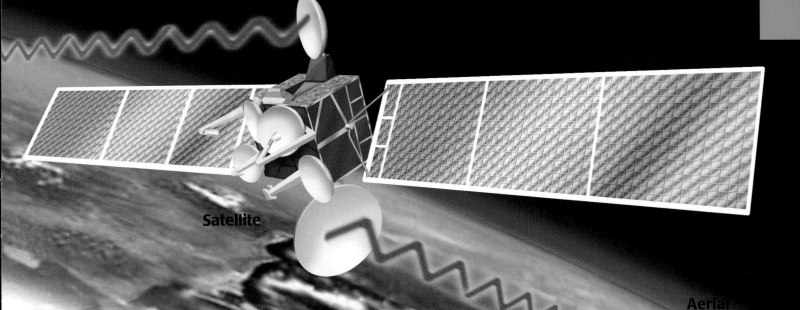

Satellite

Aerial dish

PICTURES BY SATELLITE

A TV program can be broadcast along cables or transmitted by radio signals across land (terrestrial) or through space (satellite). A satellite picks up radio signals from the ground station and retransmits them back down to aerial dishes.

An aerial dish on the roof of a house picks up the radio signals. They travel by wire to the TV receiver. It turns them back into red, green, and blue light, and sound.

Inside a TV receiver, three "guns" fire beams at the screen. They produce lines, each made up of red, green, and blue dots of light. The brightness of each dot is controlled by the color signals. Our eyes mix the lines of dots together and see them as full-color images.

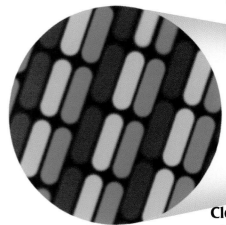

Close-up view of part of television screen

COMPUTERS

COMPUTERS are very useful electronic machines. They can be used for accessing the Internet, flying an aircraft, designing a car, storing data, and playing games.

Monitor

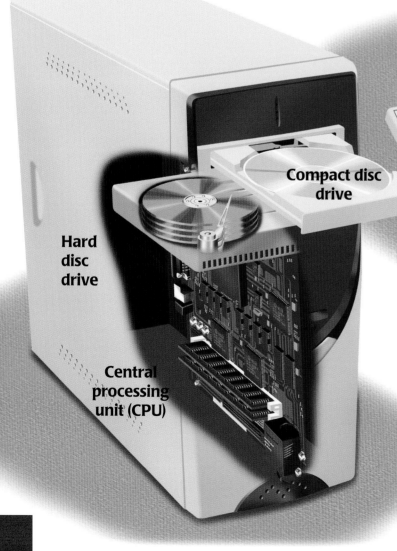

Compact disc drive

Hard disc drive

Central processing unit (CPU)

The main part of a computer is the CPU, the computer's "brain." The tiny micro-processors inside it receive instructions from a program and carry them out. Programs and data are stored in a hard disc drive. The computer's memory holds information as electrical signals.

HARDWARE AND SOFTWARE

Hardware is the basic computer equipment. It includes the central processing unit (CPU), the memory, the screen or monitor, the CD or DVD drive, keyboard, mouse, printer, and scanner. Software includes the operating system that enables the computer to work, as well as databases, games, and graphics programs.

Printer

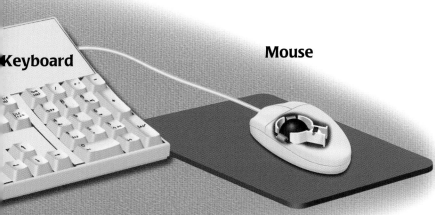

Keyboard

Mouse

A mouse is used to move the pointer on the screen. A ball inside the mouse turns a pair of wheels. Light shining through the slots in the wheels flash as the wheels turn. These flashes produce electrical signals. They tell the computer in which direction the mouse is moving.

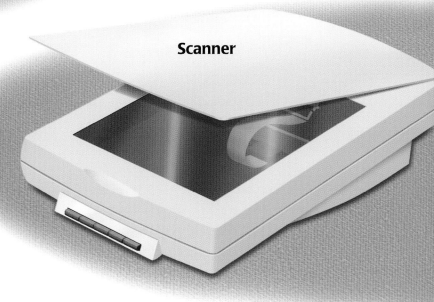

Scanner

Scanners enable pictures to be stored in a computer. A flatbed scanner works by recording the light reflected from the original picture. The light is picked up by a sensor. It divides the image into thousands of pixels that are loaded into the computer.

TELEPHONES AND THE INTERNET

TELEPHONE calls, text messages, e-mails, and computer data all travel from place to place. They are all turned into signals and sent through a vast communications network. This network is made up of telephone lines, cables, and radio and satellite links.

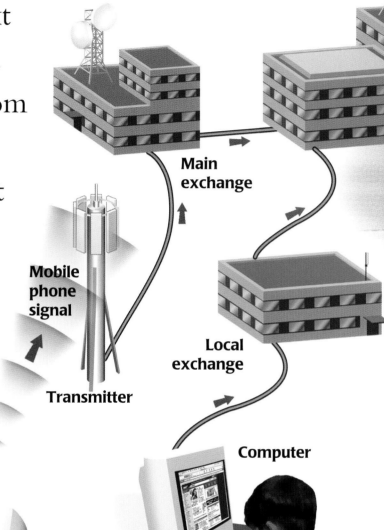

Main exchange

Mobile phone signal

Local exchange

Transmitter

Computer

Mobile phone

Mobile phones use radio waves for making calls or sending messages. The signals do not go directly from one phone to another. Instead, they are picked up by a transmitter in your local area, or "cell." They are then passed along the network to a mobile in another cell, or to an ordinary phone.

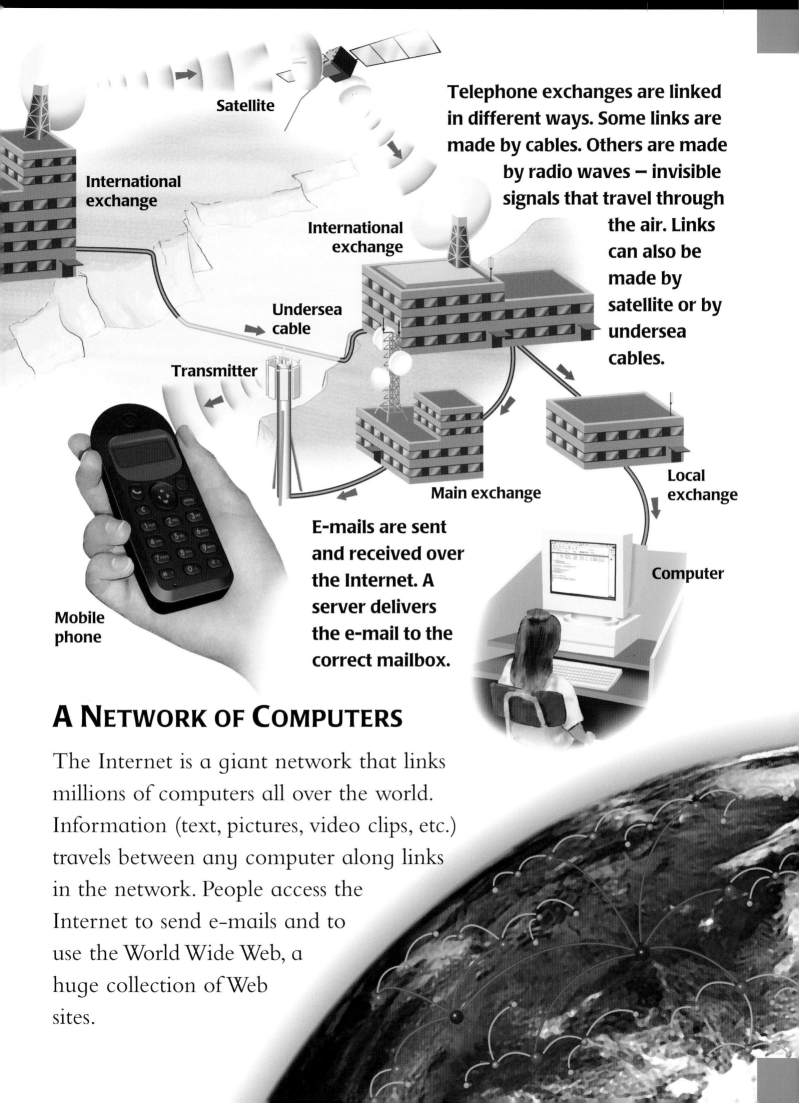

Satellite

International exchange

International exchange

Telephone exchanges are linked in different ways. Some links are made by cables. Others are made by radio waves – invisible signals that travel through the air. Links can also be made by satellite or by undersea cables.

Undersea cable

Transmitter

Local exchange

Main exchange

E-mails are sent and received over the Internet. A server delivers the e-mail to the correct mailbox.

Computer

Mobile phone

A NETWORK OF COMPUTERS

The Internet is a giant network that links millions of computers all over the world. Information (text, pictures, video clips, etc.) travels between any computer along links in the network. People access the Internet to send e-mails and to use the World Wide Web, a huge collection of Web sites.

GLOSSARY

accelerator: The gas pedal in a car.

charge-coupled device: Electronic sensor that can record images.

diesel: A type of fuel used by some cars and trains.

gravity: The natural force that draws objects toward the Earth.

microchip: A collection of circuits used to store computer programs and memory.

pixels: The smallest points of an image on a computer or digital camera.

transformer: A device that transfers electricity from one circuit to another.

INDEX